Sustainable Energy Options for Business

Philip Wolfe
WolfeWare Limited

First published in 2013 by Dō Sustainability

87 Lonsdale Road, Oxford OX2 7ET, UK

ISBN 978-1-909293-43-4 (eBook-ePub)

ISBN 978-1-909293-44-1 (eBook-PDF)

ISBN 978-1-909293-42-7 (Paperback)

A catalogue record for this title is available from the British Library.

Dō Sustainability strives for net positive social and environmental impact. See our sustainability policy at **www.dosustainability.com**.

Page design and typesetting by Alison Rayner

Cover by Becky Chilcott

For further information on Dō Sustainability, visit our website: **www.dosustainability.com**

DōShorts

Dō Sustainability is the publisher of **DōShorts**: short, high-value ebooks that distil sustainability best practice and business insights for busy, results-driven professionals. Each DōShort can be read in 90 minutes.

New and forthcoming DōShorts – stay up to date

We publish 3 to 5 new DōShorts each month. The best way to keep up to date? Sign up to our short, monthly newsletter at **www.dosustainability.com/ newsletter**. Some of our latest and forthcoming titles include:

- *Solar Photovoltaics Business Briefing*   David Thorpe
- *How to Make your Company a Recognised Sustainability Champion*   Brendan May
- *Making the Most of Standards: The Sustainability Professional's Guide*   Adrian Henriques
- *Promoting Sustainable Behaviour: A Practical Guide to What Works*   Adam Corner
- *How to Account for Sustainability: A Business Guide to Measuring and Managing*   Laura Musikanski
- *Sustainability in the Public Sector: An Essential Briefing for Stakeholders*   Sonja Powell
- *Sustainability Reporting for SMEs: Competitive Advantage Through Transparency*   Elaine Cohen
- *Sustainable Transport Fuels Business Briefing*   David Thorpe
- *The Changing Profile of Corporate Climate Change Risk* Mark Trexler & Laura Kosloff
- *The First 100 Days: How to Plan, Prioritise & Build a Sustainable Organisation*   Anne Augustine

- *The Short Guide to Sustainable Investing*   Cary Krosinsky

- *REDD+ and Business Sustainability: A Guide to Reversing Deforestation for Forward Thinking Companies*   Brian McFarland

- *Adapting to Climate Change: 2.0 Enterprise Risk Management*
Mark Trexler & Laura Kosloff

- *How Gamification Can Help Engage your Business in Sustainability*   Paula Owen

Subscriptions

In addition to individual sales of our ebooks, we now offer subscriptions. Access 60+ ebooks for the price of 6 with a personal subscription to our full e-library. Institutional subscriptions are also available for your staff or students. Visit **www.dosustainability.com/books/subscriptions** or email **veruschka@dosustainability.com**

Write for us, or suggest a DōShort

Please visit **www.dosustainability.com** for our full publishing programme. If you don't find what you need, write for us! Or suggest a DōShort on our website. We look forward to hearing from you.

Abstract

THE MAJOR ENVIRONMENTAL IMPACT of most businesses derives from their energy usage. There are many ways in which organisations can improve the sustainability of their energy supply, and doing so is likely to have a huge positive impact on environmental footprint. Happily, there is another major upside. Using energy more responsibly improves profitability, unlike many other ethical decisions which can end up increasing a business's costs.

This book introduces the options that could help make your energy usage more sustainable. After an introduction to regulatory drivers and management issues, it looks at energy opportunities in five key areas: i) saving on energy usage; ii) finding more sustainable sources of energy; iii) generating renewable electricity; iv) producing renewable heat; and v) indirect energy sustainability options. There is a checklist to help identify your best options and any quick wins, and a handy reference list, signposted from annotations in the text.

Enjoy it – this can make both the planet and the Finance Director happier.
....

About the Author

 **PHILIP WOLFE** is one of the pioneers of the renewable energy industry. He provides advisory services through WolfeWare Limited. He is also Chairman of Westmill Solar Co-operative, a panel member of the Climate Bonds Initiative, Non-Executive Director of Renewable Energy Assurance Ltd and an individual member of the Aldersgate Group.

A Cambridge first-class engineering graduate, Philip has been involved in the sustainable energy sector since the 1970s when he became the first Chief Executive of what became BP Solar. He subsequently established Intersolar Group, which from 1993 to 2002 was the sole UK manufacturer of photovoltaic cells. While Director General of the Renewable Energy Association between 2003 and 2009, he proposed the Energy Hierarchy.[D5a] He has served on the boards of industry bodies and advised government departments in the UK and Europe.

In 2012 Routledge published his book on utility-scale solar power.[A6] He has installed renewable energy in his home for heating and part of its electricity.

Contents

CONTENTS

Introduction:
Carbon and Money

THE TRIPLE BOTTOM LINE is such an elegant concept, a way for businesses to measure their positive impact on society, the environment and the bank balances of stakeholders. Elegant in theory, but all too often in practice the three objectives conflict. The most expedient materials often have the worst environmental impact; the cheapest manufacturing locations may be those with the lowest humanitarian standards.

Fortunately, there is a major aspect where the environmental and financial bottom lines are often aligned – energy. Your business's cheapest unit of energy is also the one which is least damaging to the planet – the 'negawatt', the unit you don't use.

Most business's major environmental impact comes from energy usage. This is an aspect that companies can influence directly. There are many progressive ways to decarbonise your energy use, and doing so will have a huge positive impact on your business's environmental footprint.

This book outlines the options and helps you find the most relevant ones. Of the many sustainable energy issues you could address, not all would make a significant impact. I shall try to help you identify which ones will give the biggest bang for your bucks, meet your statutory obligations and enhance your business reputation, and then summarise ways of tackling the major improvements and point to sources of support, input and advice.

What do we mean by sustainable energy?

The key objective of this book is to help you make your business's energy usage more sustainable. Usually, in doing so, you will also be able to save money; but let's come back to that.

To start with let me quickly explain what I mean by 'sustainable energy'. Fundamentally, it is an energy solution capable of enduring indefinitely. This means that there are three primary requirements:

- First, your energy supplies must be capable of sustaining your business; i.e. suitable and sufficient for its operational requirements.

- Second, the resources applied to producing your energy must be capable of lasting indefinitely. That means that commodities and materials used should be depleted no faster than they are replenished. This condition is self-evidently met for elemental renewables such as solar, hydro and wind power. It can also apply to fuel-burning energy sources such as biomass and biofuels.

- Third, sustainable energy supplies should produce no harmful by-products, including net emissions, which affect the environment; nor wastes which cannot be fully recycled. Against my definition, nuclear power, although it has relatively low emissions, would not be considered sustainable.

These principles reflect the concept of the Energy Hierarchy,[D5] showing the priorities for a sustainable approach to energy in the sequence shown in Figure 1.

FIGURE 1. The Energy Hierarchy

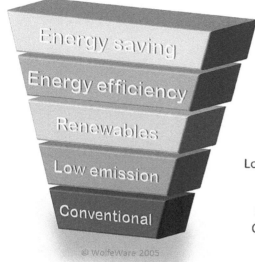

Switch off
Eliminate waste

Better appliances
Lower energy losses

Sustainable energy
production

Low carbon generation
Carbon capture

Sources of last resort
Offset to compensate

Courtesy: WolfeWare Limited

This book works down the first three levels of the hierarchy.

CHAPTER 1

Energy Trends and Policy Drivers

THESE DAYS WE TAKE ENERGY FOR GRANTED, as one of the most fundamental commodities required for modern life. It was humans' ability to harness the embedded energy in coal which fuelled the industrial revolution. The pace of industrial progress accelerated again in the early 20th century when another fossil fuel reserve – oil – made the internal combustion engine possible. The trend has continued in recent decades, with a third fossil energy resource – gas – now used for much of our heating and electricity generation.

Energy and growth

For most of this time, there has been a fairly direct link between economic growth and primary energy use. Global energy consumption has closely tracked GDP since the industrial revolution.

This relationship was of little concern to anyone for much of the 20th century. The first indication that it might not be a sustainable situation came in the 1970s when the Organisation of Petroleum Exporting Countries (OPEC), which controlled much of the world's supplies of oil and gas, unilaterally restricted output. While not catastrophic in terms of supply, this had an immediate impact on price. It also created awareness of the first constraint on energy usage – security of supply: the geographic and political mismatch between the major sources and users of energy.

The oil shortage was a temporary glitch; supplies soon reverted to normal, with consumption again rising exponentially, in turn stimulating further exploration. An enlightened few realised that these trends could not continue indefinitely, that resources must be finite and the time will come when humankind is consuming fossil fuels faster than new resources are being discovered. This is expected to happen first for petroleum (for which it might already have occurred) and hence the expression 'peak oil' was coined.

The third factor to inhibit the relentless growth in fossil fuel usage is of course climate change. Awareness of the threat of atmospheric carbon dioxide to the planet's ecosystem substantiates the un-sustainability of historical trends.

As governments have been made more forcibly aware of the triple threat of energy security, resource depletion and climatic instability, they have started to formulate policies to make energy usage more sustainable, without compromising economic growth.

The main thrusts of policy objectives have therefore been:

- to progressively decarbonise the energy resources we use; and

- to weaken the link between economic development and energy use.

There is now patchy evidence that primary energy consumption does not need to rise inexorably with economic growth. Some countries – most notably China – have started to reduce their primary energy consumption per unit of GDP.

International and national policy

A minority of states, including Germany, Japan and (perhaps surprisingly in light of their lack of involvement in the Kyoto Protocol) certain US states, have been relatively proactive in developing policies for energy efficiency and renewables. Others have tended to be more passive with much of their policy driven by international initiatives. The most notable of these have been the Kyoto Protocol of the UNFCCC[C13] and, because Europe has been a progressive force within this movement, the various environmental and energy directives from the European Union.

Here in the UK it has been the EU directives, in particular, which have driven policy in this area. In the context of this book the two most significant initiatives are the Renewable Energy Directive[B12] and the Energy Efficiency Directive.[B9] There are additionally several other European policies, affecting buildings and transport for example, which also drive towards a more sustainable energy mix. In response to these European measures, the UK has introduced a number of policies to promote energy efficiency and renewable energy.

There are also a growing number of indirect measures, in particular in relation to building standards. I do not intend to describe these in any detail because they will not affect you unless your business is in the construction sector or you are involved in building your own premises.

The UK government's Department of Energy and Climate Change (DECC) is developing specific strategies for energy efficiency[A2] and renewable energy,[A3] in the latter case supported by a so-called roadmap.[A4] DECC has rebranded the responsible sections as the Energy Efficiency Deployment Office[C7a] (EEDO) and the Office for Renewable Energy Deployment[C7b]

(ORED), respectively. The main legislative measures which support government strategies are:

> The primary energy efficiency incentive will be **The Green Deal**,[B12] being introduced in 2013. This will also absorb much of the **Energy Company Obligation**[B6] **(ECO)**, which replaced the Carbon Emissions Reduction Target[B2] (CERT) – previously the Energy Efficiency Commitment (EEC).
>
> Renewable energy has four schemes: two for electricity, the **Renewables Obligation**[B15] **(RO)** and the **Feed-in Tariffs**[B11] **(FITs)**; and one each for heat, the **Renewable Heat Incentive**[B14] **(RHI)**, and transport fuels, the **Renewable Transport Fuels Obligation**[B16] **(RTFO)**.

Beyond these energy measures (more details below), there are broader schemes aimed at reducing carbon emissions in general, primarily:

> **EU Emissions Trading Scheme**[B7] **(EUETS)** covers all major EU carbon emitters, including about 1100 UK organisations which together account for 50% of our emissions. Participants receive or buy an annual cap or 'allocation'; if their emissions exceed this level, they have to buy further allocation, but if they emit less, they can sell the surplus credits.
>
> **Climate Change Agreements**[B3] **(CCAs)** are voluntary arrangements whereby large emitters can agree emissions reduction plans in exchange for concessions again the Climate Change Levy.

> **The CRC Energy Efficiency Scheme**[B5] was introduced as an obligation on larger businesses to progressively decarbonise. For the purpose of this measure, larger businesses are those which do not fall within the EUETS or CCAs but have an annual energy consumption exceeding 6000 MWh. These businesses account for about 10% of UK emissions.

Businesses which qualify for these schemes have doubtless devised a more systematic approach to their carbon reduction strategy, but may still benefit from the proposals in this book. Businesses which currently fall below the CRC threshold should anyway be developing policies for carbon reduction. The government has indicated that it intends to progressively broaden the criteria for involvement in the CRC, so your day may be coming sooner than you think.

The following chapters pick out aspects of these measures which are of particular relevance to businesses wishing to improve their energy sustainability. Those who want to study the full national strategies and regulations in further detail can follow the links shown in the reference section at the end of this book. Try not to be put off by the number of different initiatives and the frequency with which they change.

Decarbonising national energy supplies

As part of the transition towards a more sustainable domestic energy supply, many countries have set objectives to reduce the carbon intensity of their primary energy resources. The UK's Climate Change Act[B3] sets a target of an 80% reduction by 2050.

The three largest applications of energy are transport, heating and electricity. Of the three, electricity is the one best suited to low carbon forms of generation, and so this tends to be the primary focus for carbon reduction policies.

In the UK, for example, there is a binding obligation under the European Renewable Energy Directive[B12] to increase our renewable energy contribution to 15% (from less than 5% today). The government blueprint for reaching this target by 2020[A3] envisages that renewables will supply over 30% of our electricity, but less than 10% of the country's transport fuels and heating. The government further intends that UK electricity will be 'almost completely decarbonised' by 2050[D2]. This is scheduled to be achieved mainly through renewables, but also from nuclear energy (which is classed as being carbon free) and from carbon capture and storage (CCS) in gas- and coal-fired power stations.

Policies to promote the blending of biofuels into petrol and diesel supplies and to enable biogas to be injected into the gas grid will also progressively decarbonise those energy vectors too.

So, one strategy for reducing your business carbon footprint is just to carry on regardless and rely on a gradual decarbonisation of national energy supplies to reduce your own carbon footprint. But hopefully you will want to be more pro-active than that.

CHAPTER 2

Managing your Energy Options

HAVING SEEN WHAT GOVERNMENTS are trying to do internationally and nationally to achieve a more sustainable energy mix, what can you do in your business; and how can government policies help you?

Energy strategy and management

The situation for individual companies is often a microcosm of the national picture. For most businesses, energy usage is the overwhelming part of their carbon footprint, so the most significant reduction they can make to their environmental impact is through reducing the carbon emissions caused by their energy consumption.

You need to be pro-active about your energy usage, not a victim of it. Adopt an energy management strategy. In larger businesses this means a formal policy, perhaps under ISO 50001,[D9a] supported by specific work-plans. For smaller businesses a less formal energy plan should be sufficient. In either case the senior management needs to appreciate the importance of a sustainable energy policy, and the benefits it can deliver.

What you can't influence

First, it is worth crossing off aspects where you have little influence.

Let's assume that as an independent business you have no control over the primary energy resources coming into the country. As explained above, energy sources are being gradually decarbonised, but it is unlikely that your business affects this in any meaningful way. Similarly, the energy used in transporting your products, your raw materials and your personnel around the globe are chosen by others and you probably have little influence there. You can however affect how much of this transport takes place at all, and over what distances. We will return to that later.

Next are your premises. You are probably stuck with the buildings which you use. You might be able to improve their energy usage, though a lot depends on whether you are also the owner. Most businesses do not own their premises and so the ability to influence the energy efficiency, and in many cases the supplies of energy at all, relies on the arrangements with the property landlord. We'll come back to that too.

What you can influence

So let's identify areas of energy usage that business managers can influence. We shall be looking at each of these in the following chapters of this book, and they fall broadly within four areas:

- Reducing how much energy you use

- Choosing your energy sources to improve sustainability

- Producing your own sustainable energy; and

- Compensating for unavoidable consequences of your energy usage

How to decide where to start

A purist would start at the top of the Energy Hierarchy,[D5] but if you've got a business to run, you almost certainly want to home in on aspects that improve your financial performance at the same time as your sustainability footprint. This should lead to early assessment of your energy usage and ways in which this can be reduced. Clearly, every megawatt hour your business doesn't consume not only saves money but also avoids any adverse effects from that energy generation.

The other area that can have substantial financial benefits is the production of renewable energy. It is unlikely in the UK today that renewable energy is cheaper than traditional sources, though this will change in the future. However, various incentive schemes supplement the income from renewable generation, making it in many cases more cost-effective than traditional fossil and nuclear sources.

Both of these two primary focus areas are substantially influenced by how much control you have over your energy supplies, especially in your business premises. When discussing specific opportunities therefore, the following chapters signal those which would be less suitable for businesses whose premises are held on short-term leases.

Whose job is it anyway?

I have assumed that most readers will not have been able to justify the appointment of a separate energy manager, and are themselves directors or prime movers of the business.

For a start you should include sustainable energy performance within the accountabilities of one of the senior management. He or she does not need to start out as an expert, but should have at least a rudimentary understanding of watts, amps and pennies. If you are the person who has been appointed, congratulations and enjoy it!

There are, of course, myriad consultants out there who can offer to improve the energy performance of your business. In many cases they do so without charging any fixed fee but work for a percentage of the savings they bring about. This is an area where I am happy to suspend my inherent cynicism about consultants. For many businesses that is a good way forward.

But wait a moment! Before you charge off and appoint expert advisers, why not read the rest of this book and implement some 'easy wins' yourself? If you're going to give away maybe 25% of the value of any future improvements, pick the 'low-hanging fruit' first and keep the full benefit of that to yourself.

CHAPTER 3

Using Energy Smarter

This chapter covers what many people call 'energy efficiency' or 'energy conservation'. I try to avoid those descriptions because they really cover four different concepts, and not all of them fall within the range of what you can influence in your business.

These aspects are all about accentuating the positive and eliminating the negative:

- maximising conversion efficiency of energy generation, such as the thermal efficiency of a power station or the performance ratio of a solar park;

- avoiding unneeded consumption, by switching off appliances when not in use and eliminating unproductive processes and movements;

- minimising the unproductive use of energy, for example, heat lost through poor insulation; and

- improving conversion efficiency of energy in appliances; in other words, how much energy is consumed for each useful unit of output.

The first aspect is outside your control except when applied to renewable generation, discussed in later chapters. The second is probably where you have the greatest influence, and we will return to this shortly.

Avoiding the waste of unproductive energy is another change you should be able to achieve, especially if you can make improvements to the premises you use. Finally, you can also affect the efficiency of your energy consumption by the equipment you select to use in your business.

Where's it all going?

Before you can even start, you need to know what your energy usage is now. Do you currently meter the electricity, gas and oil usage in each sector of the business and analyse usage and trends? If you do, you are already ahead of many of your peers.

If you think you have no metrics to measure your energy usage at all, however, you are almost certainly wrong. As a minimum, you get a regular bill from your electricity, gas and oil suppliers. This shows energy usage over the period and provides a starting point, even if you have nothing else.

To establish your baseline, you should audit the information that you have and look for any trends and seasonal variations, especially those which you cannot readily explain.

The next thing would be to benchmark this, to evaluate how reasonable your energy consumption is. This is not always easy and is subject to huge variations between different sectors of the economy.

The sample numbers in Figure 2 are national averages applicable to the service sector and office buildings. You should look for more specific benchmarks relevant to your business and building types; Carbon Trust[C2] and CIBSE[C4] may be able to help. If you have specific equipment or vehicles, which might be heavy energy users, you should try to assess them individually.

FIGURE 2. Simple energy benchmarks

UK average energy consumption per service employee:
■ c. 4000 kWh yr^{-1}
Offices (all electric) m^{-2}:
■ Average c. 110 kWh yr^{-1} – Good c. 75 kWh yr^{-1}

Sources: Odyssee Energy Indicators in Europe, CIBSE

Your energy audit should build up a profile of the energy usage of the business, broken down, where appropriate, between departments, buildings or work areas. If this reveals significant gaps in the knowledge base, areas of unaccountably high usage, or discrepancies between otherwise similar activities, you should get more specific data. It is relatively easy to install sub-metering equipment to get readings for individual sectors or work areas, especially in the case of electricity. Temporary clamp meters can be applied to individual distribution circuits or pieces of equipment. It shouldn't take long to find where the high usage areas are.

Once you know where the energy is going you have a start point for managing it in the future. In particular, you want to stamp on wastage and inefficient processes.

Measuring energy use should not be a one-off action. Your energy plan should include a continuous feedback loop of measuring performance, making improvements and measuring again to maximise the benefits.

Make energy visible

This is much easier if you give each of your team responsibility for the energy usage in their area. In particular, wherever possible, account for energy as a direct cost rather than an overhead. Where your production processes use electricity or heat, account for the costs within the bill of materials or cost of sales. Similarly, if energy usage can be allocated between departments, include this as part of their budget.

When energy is left as an indeterminate part of general corporate overheads, it feels like 'someone else's problem', and the opportunity to avoid wastage is lost.

The 'negawatt' – a major cash cow

Unless your business is exceptional, the biggest benefit you are likely to achieve both to your carbon footprint and financial well-being is through reduced energy usage, in particular the elimination of waste. A 'negawatt', a unit of energy you have avoided consuming, is completely clean and better than free. In the UK, every megawatt hour you stop wasting saves you about £100, and eliminates about half a tonne of carbon dioxide emissions.

The first and most obvious way of doing this is ensuring that energy consuming appliances that don't need to be on are turned off. Lighting, heating and air conditioning are perhaps the most obvious. But there are many other energy uses you may need to check. Your earlier energy audit should have identified the major ones.

This is one of the many areas where you get a real benefit from engaging the full workforce and ensuring that they understand the concept of energy sustainability. They all have similar issues at home so good practice can readily be transferred from dwellings to the workplace and vice versa. Many companies appoint an energy representative or monitor in each area of the business to oversee this and help colleagues adopt best practice.

An additional way to help ensure appliances are not left on is to use timers to regulate the hours of operation. Of course, many devices incorporate a standby mode to reduce the energy consumption, though this should not be used as an alternative to switching off unneeded appliances. Similarly, modern buildings often incorporate smart controls to reduce energy consumption at times of lower activity.

Other ways you might be warming the planet

Another, often insidiously and unnoticed, waste of energy is to allow it to seep away from its intended usage. The biggest example is poorly insulated buildings, where much of the energy, which was intended to heat the interior, ends up warming the atmosphere instead.

I do not propose to address in detail here how to improve the insulation (U-value) of your premises; there are many publications and experts on that topic. If you are spending a lot of money wasting heat in this way, you need to resolve this. If you own your premises this is relatively easy. If you don't, you should negotiate with your landlord about how improvements should be made and funded. In addition to the savings this brings to yourselves and those tenants who succeed you, it also enhances the value of the building, so the landlord should be prepared to foot a substantial part of the bill.

This type of wastage is not restricted to building insulation, and you should look for potential waste elsewhere in your business. I have seen, for example, factories where process heat was used in equipment which was so poorly lagged that much of the energy was escaping into the surrounding work areas. Even worse, the company then wasted more on air conditioning to bring the premises down to an acceptable working temperature.

Similar assessment should be applied to energy other than heat. For example, in a large area, with all the lighting on a single circuit, you may find that you are illuminating acres, but only working in one small corner.

Sustainability drive – reading the label

Energy efficiency ratings have been applied to white goods since the early 1990s; we are all familiar with the A+++ to G rating in rainbow colours. This is now applied ever more widely to other products and to buildings.

Whenever you are buying or leasing new vehicles or equipment, your specification should include demanding criteria for energy efficiency. It is important to keep these updated because standards are continually improving. What is good practice today may become mediocre next year. The average fuel efficiency for a salesman's new company car, for example, was probably less than 40 mpg in the 1990s, but should be above 50 mpg today.

The selection of vehicles may be influenced not only by the energy and emissions per mile but also by the type of fuel used, as discussed in the next chapter. Vehicles which can run on pure biofuels evidently have lower net emissions. Hybrids and gas powered vehicles also offer substantially improved sustainability and emissions ratios.

This approach should not be restricted to major capital items. When selecting any energy consuming items, such as light bulbs or flat-screen displays, the energy efficient alternative may be more expensive to buy, but is usually cheaper and less damaging in the long run.

A 7-watt LED light bulb, for example, gives as much light output as its 60-watt incandescent equivalent. Although it is also considerably more expensive, the much longer life and energy savings delivered mean that the lifecycle cost is perhaps a quarter of that of the old technology. An LED flat screen uses about 40% less energy than a conventional LCD model, itself perhaps 50% less 'hungry' than a plasma screen.

Lifecycle costing

In all your financial comparisons, therefore, it is important to look at total costs over the life of the appliance. It doesn't take long to do a quick discounted cash flow to compare the net present value or cost of each option, from the initial purchase cost, running expenses and energy usage over the life of the product.

For electric motors, for example, the purchase cost is only about 2% of the total – over 97% is the energy used over its life.

I know this is not all amazingly new and exciting. There is probably little above that you didn't already know. Hopefully, what's different this time is that you actually go and do something about it.

A number of guides on energy-saving have been published, some of which are listed in the reference section at the end of this book. DōShorts are also planning a publication specifically on energy saving.

Financial and moral support

It should now be apparent that smarter energy use offers you easy wins, which improve the sustainability of the business and also save money. Hopefully, you are even now, out on the shop floor implementing those!

Some smart energy improvements take longer to achieve a financial payback. The government's energy efficiency strategy illustrates this in the marginal abatement cost curve shown in Figure 3. The energy savings are shown along the horizontal axis, while the cost is on the vertical axis. Those below the line have a negative net cost, so in theory they pay for themselves. In practice, however, even these cost-effective approaches are often ignored, so government incentives are aimed at shortening the payback time and moving these solutions up the energy consumer's priority list.

....

FIGURE 3. Marginal abatement cost curve for UK energy efficiency

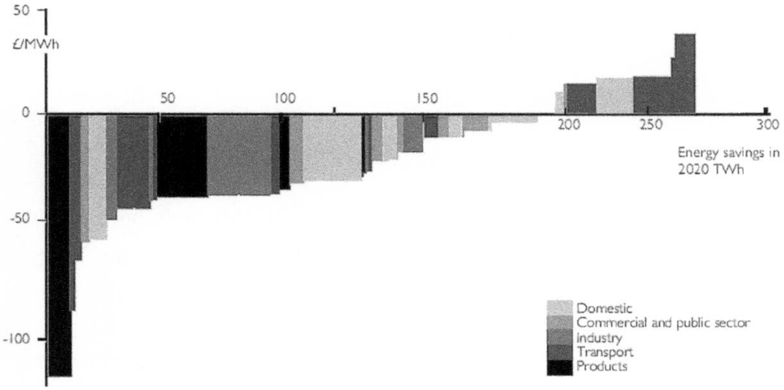

Source: Department of Energy and Climate Change
....

The main schemes that might help you are summarised below, together with references showing where further information can be obtained.

The **Green Deal**[B12] is a new scheme providing a loan for energy improvements for premises, such as insulation, heating, draught-proofing, double-glazing and some renewable energies. The loan attaches to the property (not to your business) and is paid off through electricity bills. It requires an independent assessor and the use of accredited installers.

The Green Deal is expected to absorb much of the activity undertaken by the energy suppliers under the Energy Company Obligation.[B6]

Carbon Trust may be able to help with leases, loans, finance and implementation support.[C2a]

Additionally, **Enhanced Capital Allowances**[B10] enable the cost of qualifying energy saving investments to be swiftly offset against tax.

You should also be able to find non-financial assistance from the Carbon Trust,[C2] the Energy Saving Trust,[C5] local authorities and sector-focused bodies like the Confederation of British Industry,[C3] Trades Union Congress,[C12] National Farmers Union[C10] and British Retail Consortium.[C1]

In addition to the above 'carrots' there are also some regulatory 'sticks'. Companies that fall under the scope of emissions trading schemes, for example, have to spend money buying allowances if their emissions exceed a threshold level. Similar costs will be levied in due course on organisations within the scope of the Carbon Reduction Commitment.

Additionally, you may find sustainability standards or targets being applied by your customers, particularly if they operate under an international environmental standard within the ISO 14000 family,[D8] and by local authorities and others.

CHAPTER 4

Using Greener Energy

GOOD – YOU HAVE NOW GOT ENERGY CONSUMPTION DOWN to the lowest responsible level which can support the business. The next job is to ensure that the sources of energy are as sustainable as possible.

There may be opportunities to produce your own renewable energy, and that is discussed in the next two chapters. But you will still need to buy the majority of your energy from outside suppliers; and this chapter concentrates on making those supplies as green as possible.

Fuel switching

Selecting the optimum fuel for each task is arguably an energy efficiency issue and could have been covered in the last chapter. The aim is to maximise end-to-end energy conversion efficiency for each application.

For example, electrical heaters can seem quite efficient in terms of the conversion of power to heat, but not if you consider the overall energy balance. Much of the UK's electricity comes from gas-fired power stations, converting the chemical energy of the fuel to heat. This produces steam for the turbines, which drive the generators to produce electricity. Some of that is then lost in transmission on its way to you. For each kilowatt hour of energy in the original fuel, probably less than 50% ends as heat in your premises. If instead you used a gas boiler, the efficiency should be around 90%.

If you have any vehicles, you can again choose which fuel to use against sustainability and financial criteria. The situation here is constantly changing, along with advances in engine technology mentioned above.

As a summary of the present status of sustainable transport options: pure bio-ethanol and bio-diesel fuels are available and can be considered entirely renewable. The vehicle's internal combustion engine may need to be modified to accommodate a pure biofuel, at least in the case of nominally petrol vehicles running on bio-ethanol.

Electric vehicles can be treated as being totally sustainable in fuel terms if they wholly use renewable recharging facilities, such as solar carports. The same applies to hydrogen-powered vehicles, if the hydrogen is produced entirely from renewable energy sources. Vehicles using gases such as LPG are held to have lower emissions than those powered by liquid fuels.

Much of the fuel bought on forecourts today is blended with low proportions of bio-ethanol and bio-diesel and this proportion should increase progressively as a result of the EU Renewable Energy[B13] and Biofuels Directives.[B1] Diesel emits perhaps 15% more per litre of fuel than petrol, but diesel vehicles often give better miles-per-gallon, so this choice is affected by the fuel economy of the vehicle as much as the type of fuel.

When you are changing or upgrading vehicles, you should check out the best options at that time.[A7] The situation will also change vis-à-vis electric and hydrogen-powered vehicles as the carbon intensity of the electricity in the grid progressively reduces. Gas supplies are also being made slightly cleaner, now that biogas can be injected into the gas grid under the Renewable Heat Incentive.

Renewable solid fuels, such as biomass, are covered on page 57.

Buying green power

Energy suppliers now offer various so-called 'green tariffs' for the supply of renewable energy to their customers. This is most prevalent in the electricity market, so I start with this.

You won't want a long diatribe on the structure of the UK electricity market, but I need to give an overview for the uninitiated. Since UK electricity was de-nationalised in the 1990s, there have been four classes of participants in the sector:

- generators, who produce electricity, like Drax Power, which operates the country's largest power station;

- the national transmission operator, which runs the high-voltage grid backbone and balances electricity supply to demand – this is the National Grid Company;

- the regional distribution network operators, who manage the delivery of electricity through the low voltage network to your incoming meter;

- the licensed electricity suppliers, who buy power from the generators and sell it to you and me; alongside the so-called 'big six' (who also have generation and sometimes also network operations), there are a couple of dozen smaller licensed electricity suppliers.

Because all electricity supply is through the communal grid, it is impossible to direct the output of any particular generator to any specified user. The selling of so-called green power to customers is therefore an accounting, rather than a physical, transaction. Suppliers offering renewable electricity are effectively undertaking to procure the relevant proportion of their electricity sales from renewable sources.

In many cases, suppliers charge a premium for delivering energy with a high renewable content. Some argue that this is profiteering, because all large supply companies are obliged under the Renewables Obligation to source an increasing proportion of their electricity from renewable sources in any case. The licensed supply companies also have to serve as off-takers of renewable electricity generated under the Feed-in Tariffs scheme. There is a strong body of opinion, and a logical case, that green electricity should be charged at a premium only if its production is genuinely incremental; in other words, additional to the level that has to be produced under existing statutory obligations. Because of all these complexities, there have been various initiatives to establish recognised standards for genuinely green tariffs.[D6]

Notwithstanding all these complications, a small number of electricity suppliers only buy renewable energy, and so purchasing electricity from these companies would be accepted as a fully sustainable solution. A regulatory wrinkle muddies the way in which some renewable energy is accounted for under the CRC, but let's not explore that here.

Although a small proportion of renewable gas and biofuels is now being blended into the gas network and liquid fuel supply chain respectively, there is as yet (to my knowledge) no market for pure renewable gas or transport fuels delivered in this way. It would presumably be possible to purchase and transport bio-ethanol, bio-diesel and bio-methane directly from producers. Apart from a few isolated instances in the biofuels arena, I do not believe this approach is widespread.

Finally, the purchase of renewable heat is only possible in the small minority of cases where you are connected to a heat main. Although these are few and far between, most do use largely or entirely renewable sources for heat production.

Bilateral contracts

One other option is the purchase of renewable power under bilateral contracts directly with one or more renewable energy generators. These are not entirely straightforward, because of the need to match supply and demand. Also the power you buy has to be transported through the grid network for which there is a so-called use-of-system charge. There are several energy brokers and similar organisations which specialise in this type of arrangement, but the transaction cost is probably only justified for businesses with a high electricity requirement.

CHAPTER 5

Generating Sustainable Electricity

HAVING REDUCED YOUR ENERGY consumption to a realistic minimum and ensured that the supplies of that energy are as sustainable as possible, we have hit the main issues which all businesses can address. But there is more, thanks to options which many organisations outside the energy sector may not have thought about.

In particular, there are opportunities for producing renewable energy, and there are incentives to encourage it.

The energy saving options in the previous chapters are often relatively inexpensive and pay for themselves by reducing costs or overheads. By contrast, energy generation options are typically more costly, but they too can pay for themselves, not just by reducing costs, but also by producing a new income stream for your business.

I shall start by discussing renewable electricity generation, because this is the simplest additional source to bolt on, and the one with the most widely adoptable options. Renewable heat and fuels are covered in the next chapter.

What are the renewable electricity options?

There is a surprising range of different technologies for the sustainable generation of electricity. These fall within two main classes: climatic or 'elemental' sources and fuel-powered or 'bioenergy' sources.

Some of these may not be suitable for most businesses, so this book does not consider alternatives like geothermal energy, wave and tidal power. If you happen to be situated above a hot aquifer or beside the sea, you should perhaps consider these alternatives further. The rest of us will move on.

Biomass and biofuels can generate renewable electricity through thermal conversion, and these are considered in the following chapter. That leaves three main elemental options:

Solar power

Even though it has only been properly supported since 2010, solar power is already the most widely deployed renewable energy source in the UK with over 350,000 installations. That is because the input source is daylight which is available almost everywhere. Solar is most commonly used on buildings, but other options mean you might be able to consider it, even if you do not have premises where solar panels can be mounted.

A quick overview of solar electric technology: the conversion of light to electricity in a semiconductor device is known as photovoltaics. A solar cell is a semiconductor wafer or film, treated so that when struck by photons of light, positive and negative charges are released. Conductors, usually on the front and back

---

surfaces, collect these charges and transport them to output cables to deliver power from the solar modules.

Solar cells produce power as direct current (DC) which is proportional to the area of the cell and the intensity of the light falling on it. The output voltage is a property of the materials themselves and remains fairly stable, so solar modules are designed with a number of interconnected cells to meet the output voltage required.

It is important to remember that the output is proportional to light. Heat is not required. In fact, solar cells are slightly more efficient at low temperatures, so deliver their best output on cold, bright days. The UK climate has plenty of daylight to make this technology viable. Annual light levels are similar to those in Germany, currently the world's leading market for photovoltaics. In fact, the irradiance here is only about 30% below that in Singapore on the Equator.

A typical solar system comprises an array of interconnected solar modules. These deliver their DC output to an inverter which converts it to alternating current (AC) so that it can be synchronised with the local grid. The way in which typical systems are connected means that they deliver power first to any local electrical loads. If the solar power exceeds local consumption, any surplus flows back into the grid. Similarly, if the solar generation is inadequate to meet the instantaneous on-site consumption, any shortfall is drawn from the grid.

All this is invisible to the user, who simply continues to use electricity as usual, comfortable in the knowledge that part or all of this consumption is being met by self-generated clean, renewable energy.

Wind power

The use of wind power in the UK has grown substantially in recent years, and this now produces more energy output than any other elemental renewable source (though bioenergy delivers more). Onshore wind turbines are available in capacities between about 100 watts and a few megawatts. Most offshore wind turbines are in the multi-megawatt range, but not an option we will consider here.

Wind generation technology: the fundamental ability to extract energy from the wind has been available for centuries. As the modern technology has developed it has tended to focus down towards today's standard design with three blades mounted on a horizontal axis, connected usually through a gearbox to a generator mounted in the nacelle behind the blades. There are other options available, including two-bladed devices and a variety of vertical axis designs.

In principle the power output is proportional to the swept area (i.e. the square of the blade length) and the cube of the wind speed. There is a threshold of wind speed below which the turbine ceases to turn or deliver useful output. There is also a higher wind speed above which it is unsafe for a turbine to operate and they are usually designed to slow down or stop at this level.

Wind turbines typically deliver their output in the form of AC, and are designed to synchronise with the grid either directly or through power conditioning equipment.

Hydropower

This third option, hydropower, has been part of the UK's electricity generation mix from the start, though it will be relevant to only a small minority of prospective self-generators as described below.

Fundamentally, it converts the kinetic energy of flowing water (in turn produced from the potential energy of the head – the height difference between the upper and lower water levels) into rotation of a generator.

> **Hydropower technologies:** there are several different ways in which energy can be extracted from flowing water and converted to rotating machinery. These include the Archimedes screw, the Pelton wheel, Turgo, Crossflow and Francis turbines.
>
> The selection depends on the rate of flow and the potential or head of water, and I shall not address the technicalities further here.

FIGURE 4. A hydro plant using two Archimedes Screws

Courtesy: Energy for London

As with wind turbines, hydro-generating sets are usually configured so that they can deliver a synchronised AC output directly to the grid connection.

How to choose between them

Different energy sources have different characteristics. These are usually compared using parameters such as the load factor – the proportion of time for which a generator would be expected to operate at its rated capacity. They also vary in terms of so-called despatchability. Fossil fuel power stations can in principle deliver output on demand. They store a supply of fuel and can vary the output up and down as requested by the National Grid. Within limits, the same applies to hydropower.

Solar and wind power, on the other hand, are inherently variable, delivering power only during daylight or when the wind is blowing. The times of solar output are relatively predictable; wind power is less so. Accordingly, these sources are less despatchable and have a lower load factor.

This is all very interesting to the energy expert, but should be immaterial to you, even if you choose to become a self-generator. You will remain connected to the grid so do not need to worry about the match between the times when power is being generated and consumed.

Your selection can be based therefore on which sources are practicable, and what financial returns they offer. Again this financial analysis needs to consider the full life-cycle costs and benefits as mentioned on page 33.

Hydropower is applicable only in a small minority of cases, close to a river or stream. There are several tens of thousands of old mill sites around the country and these may certainly be suitable. The two main requirements are: a reasonable flow rate and a difference of water level ideally of at least 2 m. In a few cases, where there is less head (height difference) but a fast current, it is possible to install so-called run-of-river turbines.

If you are lucky enough to have a suitable location, you will need experts to advise you, and some patience – the consenting process can be slow.

Wind power, again, is only suitable at sites with relatively high ambient wind conditions. An annual average figure of at least 5 m/s is normally considered the minimum viable wind speed. There are published maps and databases showing where the best locations might be, but wind resources are highly dependent on local topography, so you can't be sure without taking measurements at the site. This is quite easy, by installing an anemometer at the anticipated hub height of the wind turbine and taking readings, ideally for at least a year.

An additional consideration is the need for planning consent, which can be contentious especially for larger systems and in attractive countryside.

Solar systems can be deployed in many situations both building mounted and on the ground. The ideal orientation of the solar panels is tilted due south ideally at an angle between 20° and 40°. It is often possible to find roof surfaces that are suitable. However, you do not need to be too purist about the optimum orientation; the output varies only slightly with small changes of angle. Any roof faces from east, through south, to west can be considered. Equally most flat roofs are also suitable, with the solar arrays mounted on tilt structures similar to those in ground-mounted systems.

It doesn't have to be your building

Probably the most obvious option is to mount a solar system on your business premises. If you are not the property owner, the same issues on the relationship with your landlord apply as discussed in Chapter 3.

In principle you could remove a solar system and take it with you if you change premises, but you may then lose the financial incentives to be discussed later. Anyway the installation should enhance the value of the building so your landlord should be interested in participating.

Other options also exist. If you have access to land around the premises you could consider either wind turbines or ground-mounted solar installations.

Finally, of course, you don't have to use your own business location for the energy generating plant. Many companies make a living from finding and leasing suitable locations, and installing and operating renewable energy

FIGURE 5. The solar park on Westcott Airfield

Courtesy: Rockspring Property Investment Managers

plant. Do your customers, suppliers or other stakeholders have sites which they might lease to you or be interested in developing with you?

A property investment company, for example, owns a former World War II airfield, where the old runways are now unused. On one of these, they built the 1.8 MW solar park shown in Figure 5, which provides much of the power for the neighbouring industrial estate and makes a good return to the landowner.

There is a guidebook[A6] on ground-mounted solar power stations, which I can certainly recommend.

If none of the above opportunities apply to your business, that's bad luck; and you can skip to the next chapter.

What energy contribution is feasible?

When planning a new renewable electricity plant, the obvious first thought is to match its output to the energy you consume. In fact, that is not really necessary – the system's output is variable, as is your consumption – so you will stay connected to the grid. It makes more sense to match the system size to the available area, the budget you can release for the project and the returns it offers. Where the returns are good, go for the largest system you can accommodate.

Unless you have some expertise in electrical engineering and the relevant renewable energy technology, you should employ experts to manage the project implementation and to provide asset management and O&M services afterwards. When negotiating terms, be sure to link their fees to guarantees on achieving the design level of energy output. This can often be done through an agreement to share part of the income from the incentives described below. The performance and implementation of the renewable generating system will be defined by the project manager. However, I offer here for your guidance some indicative figures and a thumbnail sketch of the project development process.

Solar panels have a rated output of approximately 150 W m^{-2}. This means that a 350 m^2 barn roof could accommodate a system with a capacity of approximately 50 kW. In a good site in the Oxford area this should deliver about 50 MWh per year of useful electricity. This figure would be about 10% higher in Cornwall and maybe 15% lower in the south of Scotland.

At typical current energy prices of £0.10 p per kilowatt hour, this output has a value of £5000 per annum. The Feed-in Tariffs would deliver a further £6500 at present tariff levels (though these change regularly – see below). System costs have declined very significantly in the last 18 months. At the time of writing (and this changes fast, too) a 50 kW installation could cost under £80,000. It should be expected to have a useful life of 25 to 40 years, though the inverter at least would need replacing during this period.

Wind turbines of 50 kW capacity would have a typical tower height of 25–35 m and a blade diameter of about 20 m. In a good wind location it should deliver maybe 80 MWh per annum. Using the same assumptions as for solar power above, this power would have a value of £8000, which could be enhanced by over 200% through the Feed-in Tariffs. Wind turbine prices have been relatively stable and a machine of this size would today cost maybe £150,000 to £200,000 installed. It should have an anticipated working life of the order of 20 years, during which it will need substantial servicing of the gearbox and other moving parts.

Hydropower is highly site-specific and only an expert can make informed estimates about your particular circumstances.

Developing a renewable power project

The project development process starts with an initial estimate of the system design, detailing its capacity, the land or roof usage, and the projected annual output. In the case of wind power, this should ideally be based on actual measurements of wind speed as described above. This design allows projections of the cost and income potential of the scheme and hence a calculation of the payback period and rate of return.

Having decided to pursue the project, the next step is to obtain any required approvals. Building-mounted solar systems often fall within the scope of permitted development[B17] and so may not need planning consent. Wind turbines usually require approval, and hydro projects also need an abstraction and other licenses from the Environment Agency.[C8a]

Unless you have the reserves to finance the project on balance sheet, you will need to talk to your bank or finance providers about a loan or lease. Start this process early, so you know what their requirements are, though it may be preferable to obtain the necessary statutory consents first, as that eliminates a significant aspect of the development risk.

Once the consents and finance are in place, the detailed design can start, followed by installation and commissioning. Where you are obtaining financial support under the Feed-in Tariffs or the Renewables Obligation, it is necessary to register your plant under the scheme.

Although the fun bit is now done, it is important not to neglect suitable monitoring services to keep track of the performance of your power plant, asset management, operation and maintenance cover.

Making renewable power pay

Two government schemes have been specifically designed to provide financial support to renewable electricity generating systems.

The newer of these is the Feed-in Tariffs scheme, introduced in April 2010. This supports solar, wind and hydro schemes (plus two other technologies) up to 5 MW in capacity.

The Feed-in Tariffs[B11] **(FITs)** provide two elements of income. First, there is a generation tariff for each kilowatt hour of electricity, which the system produces, whether you use it yourself or export it to the grid. This tariff depends on the size and type of system. Tariff levels change annually (or quarterly for solar) to account for anticipated reductions in technology cost. These reductions apply only to future installations; each project's tariff is fixed at the time when the system is registered. The tariffs are then adjusted for inflation in line with the retail price index (RPI).

Second, an export tariff is paid for every kilowatt hour of power delivered to the grid. This can be negotiated bilaterally with the power off-taker, but the legislation sets a floor price which you can opt in to, currently at 4.5p/kWh for all technologies. In addition to these two income streams under the FITs, plant owners also derive a saving from any self-generated power which they use, because they buy less from their supplier.

Systems can claim FITs only if they meet various eligibility requirements. These include the need for the equipment to be certified under the Microgeneration Certification Scheme[C9] (MCS).

> Installation companies must also be accredited under the MCS and adhere to a code of conduct for customer assurance established by Renewable Energy Assurance Ltd[C11] (REAL Code).
>
> FITs tariff levels applicable to 1 May 2013 for a 50 kW generating system are 13p/kWh for solar power, 21p for wind and 19.6p for hydro.

The Feed-in Tariffs scheme is administered by the energy regulator Ofgem, which publishes details on its website,[B11a] and there are other useful online information sites.[B11b]

The other renewable power support scheme was originally developed for large-scale centralised renewable power generation:

> **The Renewables Obligation**[B15] **(RO)** was introduced in 2002 and placed a requirement on licensed electricity suppliers to procure from renewable sources a growing proportion of the electricity which they sell. In doing so they are required to produce evidence of renewable generation in the form of so-called Renewable Obligation Certificates, or ROCs.
>
> ROCs are delivered by the energy regulator Ofgem to all accredited generating stations, nominally at the rate of one ROC per megawatt-hour (MWh) produced. However, some renewable technologies receive multiple ROCs; for example, solar stations currently receive 2 ROCs/MWh, but this is scheduled to decline to 1.6 ROCs in April 2013.

> Because the supply companies need to produce ROCs as evidence of renewable generation, or pay a fine; ROCs work as the currency of the RO and have a financial value. They are traded and auctioned, recently at values around £45 each. This is equivalent to 4.5p per kilowatt hour.

The RO is available for projects over 50 kW and there is no upper limit.

Thus installations between 50 kW and 5 MW are eligible either for the RO or for the FITs. They cannot join both and have a one-off choice of which scheme to participate in, after which they cannot swap.

In principle, some renewable energy technologies can alternatively be supported under the Green Deal.[B12]

A number of guides on renewable electricity have been published, some of which are listed in the reference section at the end of this book. Dō Shorts have a publication on photovoltaics[A4] and are also planning a book on renewable power generation.

CHAPTER 6

Producing Sustainable Heat

THERE ARE ALSO MANY SUSTAINABLE heat technologies available, though perhaps fewer opportunities than for electricity, because you can apply this option only if your business uses process heat or can influence the hot water and space heating systems used in the premises.

Using renewable fuels

Before moving on to talk about installing new equipment, I should note that some existing heating systems can run on wholly or partly renewable fuels. Most oil boilers, for example, would accept fuel with a proportion of biofuels blended in. If this is what you have, check with your oil supplier to identify the most sustainable fuel you can use.

Solid fuel boilers tend to be designed for specific fuels, as further described below, but some may accept the use of biomass.

If your boiler uses natural gas from the grid, you have no real choice, though the gas supply will become progressively, if only slightly, greener as more anaerobic digestion plants feed in bio-methane under the Renewable Heat Incentive.

What are the renewable heat options?

In covering this subject, I will deal primarily with hot water and space heating for buildings. Renewable energy for process heat may well be deliverable through options such as biomass and biogas boilers, but that depends on individual circumstances and specific process requirements.

Again, renewable heat energy is available from both elemental and bioenergy sources, and I shall start of those which obtain their heat from climatic and ambient sources.

Solar heating

Solar thermal energy is used primarily for hot water, because of seasonal mismatch between sunshine availability and space heating needs.

Unlike photovoltaic solar generation, described in the last chapter, solar thermal systems are dependent on the sun's heat not its light.

Solar thermal technology: the two main alternative collector types are flat plate panels and evacuated tubes. Flat plates are effectively large radiators coloured black to maximise heat absorption. Evacuated tubes, normally made of glass, have an outer cylinder which is half mirrored to concentrate the sun's rays onto a thinner central tube, which absorbs and transfers the sun's heat. To minimise heat losses, the space between the outer and inner tubes is a vacuum.

Flat plate collectors typically operate up to 60–80°, while evacuated tubes can run well over 100°. The heat is transmitted through a collection fluid, often a mixture of water and antifreeze, and transferred to the hot water or heating system through a heat exchanger.

Heat pumps

Heat can be extracted from the ambient atmosphere using a device called an *air source heat pump*. This is rather like an air conditioning system running in reverse, and indeed many air source heat pumps are able to operate either way, in heating or cooling mode. Because the heat pump acts effectively as a heat amplifier, it can provide its heat output at temperatures significantly above that of the ambient air.

The device draws in air, extracts some of its heat, and returns the air to the atmosphere a few degrees cooler. The heat is transferred to a fluid for use in the building. This is often a hot water circuit, but air-to-air heat pumps output directly to warm air heating systems.

Heat pumps generally deliver their output at lower temperature than fuel burning boilers but, unlike the latter, operate continuously. This gives a more uniform heating profile – lower temperatures over a longer period. Under-floor heating is considered to be the optimal delivery approach for heat pumps, but they can also supply warm air and traditional hot water systems (though radiators may need to be re-sized).

Heat pump technology: to keep it simple, without going into the physics, it is easiest to think of a heat pump as a back-to-front fridge. A compressor pressurises a refrigerant gas until it liquefies, causing the refrigerant to heat up; this energy is extracted in a heat exchanger to warm the building. The refrigerant is then allowed to expand through a valve, which reduces its pressure, so it cools down and reverts to gas, which is passed through a second heat exchanger to be warmed by the ambient air (or other heat source).

This cycle is repeated continuously.

The pump is driven by electricity so it does require an external energy input, but the heat delivered is a multiple of the electrical energy used. This parameter is known as the coefficient of performance (COP). If a heat pump operates with a COP of 4, that means that for every kilowatt hour of electricity consumed, 4 kWh of heat are delivered.

A similar principle is applied in two other types of heat pump:

Water source heat pumps, rather than taking the heat from the air, use a body of water, for example, a neighbouring pond, river or lake. Provided that the water volume is sufficient, ideally with some through-flow, this can be a very efficient renewable heat source.

Ground source heat pumps use a similar principle to extract heat from the earth. A heat collection fluid of water with antifreeze is pumped through a ground loop and absorbs heat from the surrounding soil or rock. The ground loop comprises either a series of vertical boreholes or horizontal coils or panels, buried (usually about 2 m) below ground over a wide area.

This technology is different from deep geothermal energy which extracts heat from the warm core of the earth. As in the previous chapter this is not covered here because it is very location specific.

Bioenergy heat sources

The other group of renewable heat options are based on bioenergy where biomass or biofuels are burnt to produce heat. Bioenergy is considered sustainable, unlike fossil fuels, because it is derived from grown feedstock. During the growing cycle these biomass crops absorb from the atmosphere at least as much carbon dioxide as is released when they are later burnt, so there are no net emissions over the fuel cycle.

Many crops and waste streams can be used for bioenergy; some are specially grown, and some by-products from other activities. Rather than describe all these feedstocks and conversion technologies (apart from anaerobic digestion below), suffice it to say that the output fuel ends up as biogas, solid biomass or a liquid biofuel.

As an energy user, our interest is in converting this fuel to heat. Boilers for biogas or bio-liquid fuels are often standard units, though the burners and other parts may need to be modified to accommodate biofuels.

Biomass heating systems

A solid biomass boiler has some notable differences from other solid fuel systems. Biomass boilers are designed for one of three fuel types: wood chips, pellets or logs (often called solid roundwood). In addition to the boiler itself, storage is needed to suit the selected fuel type.

Chips and pellets have the advantage that they can be delivered into the boiler using automated fuel delivery systems, requiring very little manual intervention. Similarly, many modern biomass boiler systems incorporate semi-automated ash extraction to minimise maintenance.

> **Biomass boilers** are operationally the same as a traditional solid fuel boiler. The fuel is burnt and the heat is delivered to the space heating and hot water system of the premises, usually through a hot water loop.
>
> The combustion process needs to be controlled to be thermally efficient and to minimise emissions up the flue to the atmosphere. Not all biomass fuels and boilers meet demanding clean air standards in some parts of the country; this is an important matter to check with the supplier.

Anaerobic digestion

Anaerobic digestion (AD) is used to produce bio-methane gas from biomass sources. It can be tailored to a wide range of different feedstocks from custom crops like maize, agricultural by-products including straw, or recycling sources such as food waste.

> **Anaerobic digestion** is a natural process where, in the absence of oxygen, micro-organisms break down organic matter into bio-methane gas and a slurry by-product, which can be used as fertiliser.
>
> The process takes a few days in large covered digestion tanks, with the bio-methane gas being extracted at the top.

Combined heat and power

When any of these bioenergy fuels are used at sufficient scale to produce heat, a further option is to install a combined heat and power (CHP) plant, which uses part of the heat output to drive a generator and produce electricity at the same time.

For any installation over 50 kW, this electricity output would be eligible for support under the Renewables Obligation, described in the previous chapter. At present, the RO allocates an additional premium of ½ ROC per MWh for CHP plants. In the case of anaerobic digestion plants under 5 MW, the electrical output would alternately be eligible for support under the Feed-in Tariffs also described above.

Choosing your renewable heat system

Again, let's dispense quickly with the options that are not widely applicable. Water source heat pumps are appropriate only if you have a suitable body of water nearby. Anaerobic digestion needs a regular source of suitable feedstock, most likely to be available in the agricultural and food processing sectors.

The other options can all be considered for space heating and hot water.

Solar heating systems are best mounted on a suitable roof, with similar requirements to those for solar electric systems in the previous chapter.

Heat pumps should be considered for new buildings and those off the gas grid. *Air source* heat pumps are much like large air-conditioners so need a suitable external mounting location where their moderate noise output will not be distracting. *Ground source* heat pumps with horizontal ground loops need a substantial land area to bury these in, as a very rough rule-of-thumb maybe four times the floor area of the heated premises. Vertical boreholes only need a fraction of this area, but it must be suitable for the drilling of an array of boreholes spaced ideally at least 10 m apart.

Biomass boilers can often be housed in a similar location to a traditional heating system, with a fuel store close by and accessible for deliveries. Fuel suppliers often deliver chip and pellets by blowing them through a tube. The volume required for the fuel store, depending on the frequency of deliveries, might be 10–20 cubic metres for a 50 kW boiler. Be sure to find a suitable fuel supplier before committing to this technology.

What energy contribution is feasible?

Except for solar hot water systems, renewable heating is usually sized to meet actual thermal requirements (unlike the approach for renewable power systems above). Again, you will almost certainly need to retain experts to design and implement your chosen solution.

Solar hot water systems are backed up by other heat sources and are usually designed to meet only part of the heat needed. Typical household systems use panels totalling about 4 m^2 at typical costs around £4800, so assume you will need some multiple of this, depending on the occupancy of the building. Even with financial support, as described below, the payback for such a system will be several years.

The other space heating technologies described above would often be installed similar to a traditional boiler. Though often somewhat larger in size, there should be few particular installation issues, apart from fuel storage or ground loops, as described above.

It is not particularly helpful to generalise about the size and performance of these heating systems, as each would be tailored to the individual requirements. Although system costs vary depending on the technology, the support under the Renewable Heat Incentive has been designed to give similar returns for each of these options. The financial issues are therefore discussed under the RHI below. In the case of renewable heating, your decision can be based heavily on which technology best suits your situation and requirements.

Making renewable heat pay

The UK government was one of the first to introduce a comprehensive support scheme for renewable heat. The Renewable Heat Incentive[B14] was introduced for non-domestic users in 2011 and is due to be extended to domestic heating in 2013. It also offers a tariff for injecting bio-methane from AD plants into the gas grid.

> **The Renewable Heat Incentive**[B14] **(RHI)** is a companion measure to the Feed-in Tariffs and was introduced under the same primary legislation, the 2008 Energy Act.[B8] It is similar to the FITs in that it sets specific tariffs for each kWh of heat produced depending on technology and system size. Not all renewable heat sources are eligible.
>
> For business users, the heat output must be measured using a qualifying calibrated meter. For large systems, and those in multiple buildings, there are more demanding metering requirements and an independent expert must be retained. Other safeguards ensure

that are only useful heat is compensated. As for the FITs, users are protected by the requirement for installers to be MCS-accredited and REAL Code participants.

Sample tariff levels currently applicable for a 50 kW heating system for non-domestic use are 8.5p/kWh for solar hot water, 4.5p for ground source heat pumps and 7.9p for biomass boilers.

The RHI is administered by the energy regulator Ofgem, which publishes details on its website,[B14a] and there are other useful online information sites.[B14b] The government funds the RHI through Ofgem, unlike the FITs, which are paid by electricity consumers through the licensed suppliers.

When setting tariff levels, the government's intention was to ensure that all technologies, apart from solar thermal, deliver a financial return on investment of about 12%. This means simplistically that investments in qualifying renewable heat plants should pay back in about eight years.

Some renewable energy technologies can alternatively be supported under the Green Deal.[B12]

CHAPTER 7

Indirect Measures

HAVING COVERED THE MAIN ASPECTS which you can actively address to improve the sustainability of energy in your business, let's touch finally on a few indirect approaches.

Waste to energy

Like energy, the waste sector is subject to sustainability drivers. The waste hierarchy encourages recycling; and one way in which petroleum-based and biomass waste streams can be reused is by burning them for energy.

Although this may already be happening to the waste from your business, perhaps you could be more proactive about finding outlets for any wood, plastic, agricultural or food wastes. The combined effects of landfill tax and other constraints on the waste sector with renewable energy incentives means that many suitable waste streams have a positive financial value. In that case you could get paid for having it taken away, at the same time ensuring there is no danger of this waste going to landfill.

Smarter ways to get there

The fuel options for your own vehicles were discussed in Chapter 4. You should also consider the energy footprint of external transport options. Typical energy intensity per load mile is lower on trains or public transport and highest in aviation. When deciding between different alternatives, this should be taken into account as well as the financial cost.

Finding the local sources for your raw materials reduces the energy required to transport supplies to your premises.

You can encourage your staff to adopt less energy intensive transport options themselves. Promote and facilitate car-sharing. Provide secure cycle storage, showers and changing rooms in your premises, so those living nearby can cycle to work instead of driving.

It's all about people

The people in your business will be fundamental to delivering the energy savings discussed in Chapter 3. It may be that you can make an additional contribution to energy sustainability by helping them to implement similar improvements in their own homes. One easy way of doing this is by making simple information on energy saving and renewable energy available to your employees, maybe through your intranet to save on paper. The Energy Saving Trust[C5] and others provide such materials.

You could consider temporary or permanent financial assistance to help employees adopt sustainable energy measures. Installing renewables at home, with support from the FITs or RHI, offers attractive financial returns for your people. However, they may have trouble finding the initial capital cost required. Could your business make loans to employees to be re-paid progressively from the tariff income they receive?

You can also save energy by the working practices your business adopts. Each day a member of staff works from home, he or she saves the energy, and the cost, which would have been used getting to work. Obviously you need to ensure that productivity is not compromised.

Working with suppliers and other stakeholders

There may, similarly, be opportunities for your business to have a positive influence on the energy sustainability of other people and organisations with which you do business. As a minimum, you should ask all your suppliers to provide details of their sustainable energy policy. If they have good practices, there may be useful ideas you can borrow. If they have no concern for the issue at all, you may want to find alternative suppliers – or encourage them to buy this book!

Once you have followed all these suggestions, your business should be using very much less unsustainable energy than before. To the extent that there remains an unavoidable environmental and carbon footprint from your energy usage, you might wish to compensate for this in other ways.

There are, for example, schemes for offsetting a specific level of carbon emissions by paying for emission reductions elsewhere. I am somewhat sceptical about some of these schemes, being unconvinced that all are genuinely incremental. However, standards[D1] are being developed in this area, and it should certainly be considered.

Finally, if your business is fortunate enough to have significant cash reserves, there are opportunities for these to be profitably invested in other sustainable energy projects. Many specialist funds now support renewable energy and energy efficiency projects. There is also a growth in so-called Green Bonds,[D3] a new form of financial instrument designed to support similar projects.

. . . .

Conclusions:
Evaluating your Options,
Getting Some Quick Wins

HOPEFULLY, THERE ARE SEVERAL IDEAS in this book that make sense in your business. But with other things on your plate, you can't do everything.

To help shortlist the first things to act on, do a quick assessment of which ideas could be most important for you, by preparing a matrix showing:

- The action points (from Table 1 below)

- What that aspect is worth or costing today (kWh or £/year)

- Rank its significance to your business (see last column in Table 1; those with an asterisk are less suitable if you have no control over your premises). Rank from 0 to 10, say

- Then give it an action category; for example: A = Act now, B = standBy actions, C = Consider later, D = Done, E = irrElevant

Only have three in category A at any one time. When each one is done, bring forward the next one from the B- or C-list.

TABLE 1. Grading criteria for energy vectors and actions

Page	Action point	More significant
23	Energy policy & strategy	Larger companies
23	Energy plan	Smaller companies
25	Management responsibility	More employees or locations
26	Energy consultants	After the big wins
28	Measure energy use	Always
28	Additional metering	Energy-intensive processes/areas
28	Benchmark against others	Always
30	Allocate to product costs	Manufacturers and producers
30	Allocate to departments	Multiple departments/work areas
30	Policies to 'switch off'	Always
31	Appliances with 'stand-by'	Always
31	Appoint energy champions	Multiple work areas
31	'Smart' controls	* New buildings
31	Building energy efficiency	* Older buildings
32	Insulate heated plant	Significant process heat
32	More switches in large area	E.g. warehousing
32	Energy specs for plant	New equipment purchase
32	Energy specs for vehicles	Own vehicle fleet
33	Policies for small purchases	Always: light bulbs, computers, etc.
33	Lifecycle costing	New equipment purchase
35	Green Deal support	* Improve efficiency of buildings
35	Assistance for financing	New energy efficiency purchases
35	Capital allowances	New energy efficiency purchases
37	Consider gas for heating	* Existing electrical heating
38	Optimise vehicle fuels	Own vehicle fleet

Page	Action point	More significant
38	Optimise vehicle purchases	Own vehicle fleet
39	Consider green tariff	Significant electricity purchases
40	Buying greener gas/oil	Direct gas/oil deliveries
41	Direct electricity purchase	Major electricity user
44	Geothermal power option?	Above hot aquifer
44	Tidal power option?	Beside the sea
48	Consider hydropower	Adjacent to mill or stream
49	Consider wind power	Suitable high-wind location
49	Consider solar power	* Southerly roof areas available
49	Off-building systems	Suitable land available
50	Other people's land	Suitable suppliers/partners
53	Renewable power project	'Yes' to any of the last seven points
54	Support from FITs	Qualifying project under 5 MW
55	Support from RO	Qualifying project over 50 kW
57	Biofuels for your oil boiler?	Can (be modified to) accept biofuels
57	Biogas for your gas boiler?	Biogas source available
62	Consider AD	Agricultural/food wastes available
63	Consider solar heating	* Southerly roof areas available
63	Consider air-source heat	* Suitable external location
63	Ground source heat	* Suitable land area
63	Water source heat	* Neighbouring water source
64	Consider biomass heat	* Suitable fuel store area
64	Renewable heat project	'Yes' to any of the last six points
65	Support from RHI	Qualifying renewable selected
67	Selling energy wastes	Wood, food or bioenergy wastes
67	Best business travel	Frequent business trips

Page	Action point	More significant
68	Local sourcing	High volume of supplies
68	Cycle facilities	Many local employees
68	Energy info for employees	Employees are home-owners
68	Help for home energy	Staff interested in renewables
68	Home-working policy	Phone- or computer-based staff
69	Copy energy best practice	Suppliers have good practices
69	Supplier selection	Suppliers have bad practices
71	Draw up action list	Always

Off you go then – what are you waiting for?

Bibliography, Links & References

A. Other publications worth reading

A1. Carbon Trust:

    a. Employee awareness and office energy efficiency guides: http://www.carbontrust.com/resources/guides/energy-efficiency/employee-awareness-and-office-energy-efficiency

    b. Energy management for business guide: http://www.carbontrust.com/resources/guides/energy-efficiency/energy-management

A2. Energy Efficiency Strategy: *The Energy Efficiency Opportunity in the UK*, Department of Energy and Climate Change, 2012: http://www.decc.gov.uk/en/content/cms/tackling/saving_energy/what_doing/eedo/eedo.aspx

    a. *Factors influencing energy behaviours and decision-making in the non-domestic sector*, CSE[C6] et al., 2012: http://www.decc.gov.uk/assets/decc/11/tackling-climate-change/saving-energy-co2/6925-what-are-the-factors-influencing-energy-behaviours.pdf

A3. Renewable Energy Strategy: *UK National Renewable Energy Action Plan*, Department of Energy and Climate Change, 2010: http://www.decc.gov.uk/en/content/cms/meeting_energy/renewable_ener/uk_action_plan/uk_action_plan.aspx

A4. *UK Renewable Energy Roadmap*, Department of Energy and Climate Change, 2011 (updated 2012): **http://www.decc.gov. uk/en/content/cms/meeting_energy/renewable_ener/re_ roadmap/re_roadmap.aspx**

A5. Thorpe, D. 2012. *Solar Photovoltaics Business Briefing* (Oxford: Dō Sustainability): **http://www.dosustainability.com/shop/solar-photovoltaics-business-briefing-p-3.html**

A6. Wolfe, P. 2012. *Solar Photovoltaic Projects in the Mainstream Power Market* (London: Routledge): **http://www.routledge.com/ books/details/9780415520485/**

A7. Thorpe, D. 2012. *Sustainable Transport Fuels Business Briefing* (Oxford: Dō Sustainability): **http://www.dosustainability.com/ shop/sustainable-transport-fuels-business-briefing-p-8.html**

B. Regulatory incentives and legislation

B1. (EU) Biofuels Directive 2009: Directive 2009/28/EC, European Commission, Brussels: **http://eur-lex.europa.eu/LexUriServ/ LexUriServ.do?uri=Oj:L:2009:140:0016:0062:en:PDF**

B2. Carbon Emission Reduction Target (CERT) – formerly Energy Efficiency Commitment (EEC). *Closed at the end of 2012 to make way for the Green Deal:* **http://www.decc.gov.uk/en/content/ cms/funding/funding_ops/cert/cert.aspx**

B3. (UK) Climate Change Act 2008: **http://www.legislation.gov.uk/ ukpga/2008/27/contents**

B4. Climate Change Agreements (CCA): **http://www.decc.gov.uk/en/ content/cms/emissions/ccas/ccas.aspx**

B5. CRC Energy Efficiency Scheme (formerly *Carbon Reduction Commitment*): http://www.decc.gov.uk/en/content/cms/emissions/crc_efficiency/crc_efficiency.aspx

B6. Energy Company Obligation (ECO): https://www.gov.uk/energy-company-obligation

B7. EU Emissions Trading Scheme (EUETS): http://www.decc.gov.uk/en/content/cms/emissions/eu_ets/eu_ets.aspx

B8. (UK) Energy Act 2008: http://www.legislation.gov.uk/ukpga/2008/32/contents

B9. (EU) Energy Efficiency Directive 2012: Directive 2012/27/EU, European Commission, Brussels: http://eur-lex.europa.eu/JOHtml.do?uri=OJ:L:2012:315:SOM:EN:HTML

B10. Enhanced Capital Allowances: http://etl.decc.gov.uk/etl

B11. Feed-in Tariffs (FITs): http://www.decc.gov.uk/en/content/cms/meeting_energy/renewable_ener/feedin_tariff/feedin_tariff.aspx

   a. Guidance from the scheme administrator, Ofgem: http://www.ofgem.gov.uk/Sustainability/Environment/FITs/Pages/FITs.aspx

   b. Useful information website: http://www.fitariffs.co.uk/

B12. The Green Deal: https://www.gov.uk/green-deal-energy-saving-measures

B13. (EU) Renewable Energy Directive 2009: Directive 2009/28/EC, European Commission, Brussels: http://eur-lex.europa.eu/LexUriServ/LexUriServ.do?uri=OJ:L:2009:140:0016:0062:en:pdf

B14. Renewable Heat Incentive (RHI): http://www.decc.gov.uk/en/content/cms/meeting_energy/renewable_ener/incentive/incentive.aspx

a. Guidance from the scheme administrator, Ofgem:
http://www.ofgem.gov.uk/e-serve/RHI/Pages/RHI.aspx

b. Useful information website: http://www.rhincentive.co.uk/

B15. Renewables Obligation (RO): http://www.decc.gov.uk/en/
content/cms/meeting_energy/renewable_ener/renew_obs/
renew_obs.aspx

B16. Renewable Transport Fuels Obligation (RTFO): https://www.gov.
uk/renewable-transport-fuels-obligation

B17. The Town and Country Planning (General Permitted Development)
(Amendment) (England) Order 2012: http://www.legislation.gov.
uk/uksi/2012/748/made

C. Other useful organisations and websites

C1. British Retail Consortium; A better retailing climate: http://www.
brc.org.uk/brc_policy_content.asp?iCat=43&iSubCat=673&spoli
cy=Environment&sSubPolicy=A+Better+Retailing+Climate

C2. Carbon Trust: http://www.carbontrust.com/home

a. Energy Efficiency Finance and Implementation support:
http://www.carbontrust.com/client-services/technology/
implementation

C3. CBI; Energy and Climate Change: http://www.cbi.org.uk/business-
issues/energy-and-climate-change/

C4. Chartered Institution of Building Services Engineers (CIBSE):
https://www.cibse.org/

C5. Energy Saving Trust (primarily focused on sustainable energy for
individual consumers): http://www.energysavingtrust.org.uk/

C6. Centre for Sustainable Energy (CSE): http://www.cse.org.uk/

C7. Department of Energy and Climate Change

    a. Energy Efficiency Deployment Office (EEDO): http://www.decc. gov.uk/en/content/cms/tackling/saving_energy/what_doing/ eedo/eedo.aspx

    b. Office for Renewable Energy Deployment (ORED): http://www. decc.gov.uk/en/content/cms/meeting_energy/renewable_ ener/ored/ored.aspx

C8. Environment Agency

    a. Hydropower regulation: http://www.environment-agency.gov.uk/ business/topics/water/32022.aspx

C9. Microgeneration Certification Scheme (MCS): http://www.microgenerationcertification.org/

C10. National Farmers Union; Farm Energy Service: http://www.nfufarmenergyservice.com/about-our-service/about-our-service/

C11. Renewable Energy Assurance Limited (REAL Code): http://www.realassurance.org.uk/

C12. Trades Union Congress; Green Workplaces Network: http://www.tuc.org.uk/workplace/index.cfm?mins=392&minors =87&majorsubjectID=2

C13. UN Framework Committee on Climate Change (UNFCCC): http://www.unfccc.int/

D. Further references in the text

D1. Carbon Offset standards: **http://www.carbonneutral.com/ knowledge-centre/offsetting-explained/standards/**

D2. The Carbon Plan: *Delivering Our Low Carbon Future*, HM Government, 2011: **http://www.decc.gov.uk/assets/decc/11/ tackling-climate-change/carbon-plan/3702-the-carbon-plandelivering-our-low-carbon-future.pdf**

D3. The Climate Bonds Initiative: **http://climatebonds.net/**

D4. *Energy Benchmarks for Public Sector Buildings in Northern Ireland*, Chartered Institution of Building Services Engineers: **https://www.cibse.org/pdfs/energy_benchmarks.pdf**

D5. The Energy Hierarchy was first postulated by the author:

a. Philip Wolfe, *A Proposed Energy Hierarchy*, WolfeWare, 2005: **http://www.wolfeware.co.uk/Documents/Reports/ EnergyHierarchy.pdf**

b. Endorsed in *The Sustainable Energy Manifesto*, Renewable Energy Association et al., 2006; updated 2008: **http://www. wolfeware.co.uk/Documents/Reports/Energy2020Manifesto. pdf**

c. Later refined in *The Energy Hierarchy*, Institution of Mechanical Engineers, 2009: **http://www.imeche.org/Libraries/Position_ Statements-Energy/EnergyHierarchyIMechEPolicy.sflb.ashx**

D6. Green Energy Scheme: **http://www.greenenergyscheme.org/**

D7. ISO 9000: Quality Management – family of standards, International Standards Organisation: **http://www.iso.org/iso/ iso_9000**

D8. ISO 14000: Environmental Management – family of standards, International Standards Organisation: http://www.iso.org/iso/home/standards/management-standards/iso14000.htm

D9. ISO 50001: Energy Management standard, International Standards Organisation: http://www.iso.org/iso/home/standards/management-standards/iso50001.htm

   a. Energy management policy can alternatively be formalised as part of a management system under either of the other standards families[D7,D8] above.

   b. See also: Henriques, A. 2012. *Making the Most of Standards: The Sustainability Professional's Guide* (Oxford: Dō Sustainability): http://www.dosustainability.com/shop/making-the-most-of-standards-the-sustainability-professionals-guide-p-12.html

D10. World Energy Efficiency Indicators, World Energy Council: http://wec-indicators.enerdata.eu/

For Product Safety Concerns and Information please contact our EU
representative GPSR@taylorandfrancis.com
Taylor & Francis Verlag GmbH, Kaufingerstraße 24, 80331 München, Germany